Review of the Need for a Large-scale Test Facility for Research on the Effects of Extreme Winds on Structures

Committee to Review the Need for a Large-scale Test Facility for Research on the Effects of
Extreme Winds on Structures
Board on Infrastructure and the Constructed Environment
Commission on Engineering and Technical Systems
National Research Council

NATIONAL ACADEMY PRESS
WASHINGTON, D.C. 1999

NOTICE: The project that is the subject of this report was approved by the Governing Board of the National Research Council, whose members are drawn from the councils of the National Academy of Sciences, the National Academy of Engineering, and the Institute of Medicine. The members of the committee responsible for the report were chosen for their special competencies and with regard for appropriate balance.

This report has been reviewed by a group other than the authors according to procedures approved by a Report Review Committee consisting of members of the National Academy of Sciences, the National Academy of Engineering, and the Institute of Medicine.

The National Academy of Sciences is a private, nonprofit, self-perpetuating society of distinguished scholars engaged in scientific and engineering research, dedicated to the furtherance of science and technology and to their use for the general welfare. Upon the authority of the charter granted to it by the Congress in 1863, the Academy has a mandate that requires it to advise the federal government on scientific and technical matters. Dr. Bruce Alberts is president of the National Academy of Sciences.

The National Academy of Engineering was established in 1964, under the charter of the National Academy of Sciences, as a parallel organization of outstanding engineers. It is autonomous in its administration and in the selection of its members, sharing with the National Academy of Sciences the responsibility for advising the federal government. The National Academy of Engineering also sponsors engineering programs aimed at meeting national needs, encourages education and research, and recognizes the superior achievements of engineers. Dr. William Wulf is president of the National Academy of Engineering.

The Institute of Medicine was established in 1970 by the National Academy of Sciences to secure the services of eminent members of appropriate professions in the examination of policy matters pertaining to the health of the public. The Institute acts under the responsibility given to the National Academy of Sciences by its congressional charter to be an adviser to the federal government and, upon its own initiative, to identify issues of medical care, research, and education. Dr. Kenneth I. Shine is President of the Institute of Medicine.

The National Research Council was organized by the National Academy of Sciences in 1916 to associate the broad community of science and technology with the Academy's purposes of furthering knowledge and advising the federal government. Functioning in accordance with general policies determined by the Academy, the Council has become the principal operating agency of both the National Academy of Sciences and the National Academy of Engineering in providing services to the government, the public, and the scientific and engineering communities. The Council is administered jointly by both Academies and the Institute of Medicine. Dr. Bruce Alberts and Dr. William Wulf are chairman and vice chairman, respectively, of the National Research Council.

This study was supported by Grant No. DE-FG07-98ID13722 from the U.S. Department of Energy to the National Academy of Sciences. Any opinions, findings, conclusions, or recommendations expressed in this publication are those of the authors(s) and do not necessarily reflect the view of the organization or agency that provided support for this project.

International Standard Book Number 0-309-06483-X

Available in limited supply from: Board on Infrastructure and the Constructed Environment, 2101 Constitution Avenue, N.W., HA 274, Washington, D.C. 20418, (202) 334-3376

Additional copies of this report are available for sale from: National Academy Press 2101 Constitution Avenue, N.W. Washington, DC 20418 Call 800-624-6242 or 202-334-3313 (in the Washington Metropolitan Area). This report is also available on line at **http://www.nap.edu**

Printed in the United States of America.

Copyright 1999 by the National Academy of Sciences. All rights reserved.

Committee to Review the Need for a Large-scale Test Facility for Research on the Effects of Extreme Winds on Structures

JACK E. CERMAK, *chair*, Colorado State University, Fort Collins
ALAN G. DAVENPORT, University of Western Ontario, London
MICHAEL P. GAUS, State University of New York at Buffalo
STEPHEN R. HOOVER, Kemper/NATLSCO, Long Grove, Illinois
NICHOLAS P. JONES, Johns Hopkins University, Baltimore, Maryland
AHSAN KAREEM, University of Notre Dame, Notre Dame, Indiana
RICHARD J. KRISTIE, Wiss, Janey, Elstner Associates, Inc., Northbrook, Illinois
WILLIAM F. MARCUSON, III, U.S. Army Corps of Engineers, Vicksburg, Mississippi
JOSEPH E. MINOR, University of Missouri-Rolla
JOSEPH PENZIEN, International Civil Engineering Consultants, Inc., Berkeley, California
MARK D. POWELL, National Atmospheric and Oceanic Administration, Miami, Florida
TIMOTHY A. REINHOLD, Clemson University, Clemson, South Carolina
ELEONORA SABADELL, National Science Foundation, Arlington, Virginia
EMIL SIMIU, National Institute of Standards and Technology, Gaithersburg, Maryland

Staff

RICHARD G. LITTLE, Study Director
MICHELLE L. PORTERFIELD, Consultant
JENIFER BOLSER, Project Assistant
AMANDA PICHA, Project Assistant

Board on Infrastructure and the Constructed Environment

JAMES O. JIRSA, *chair*, University of Texas, Austin
BRENDA MYERS BOHLKE, Parsons Brinckerhoff, Inc., Herndon, Virginia
JACK E. BUFFINGTON, University of Arkansas, Fayetteville
RICHARD DATTNER, Richard Dattner Architect, P.C., New York, New York
CLAIRE FELBINGER, American University, Washington, D.C.
AMY GLASMEIER, Pennsylvania State University, University Park
CHRISTOPHER M. GORDON, Massachusetts Port Authority, Boston
NEIL GRIGG, Colorado State University, Fort Collins
DELON HAMPTON, Delon Hampton & Associates, Washington, D.C.
GEORGE D. LEAL, Dames & Moore, Inc., Los Angeles, California
VIVIAN LOFTNESS, Carnegie Mellon University, Pittsburgh, Pennsylvania
MARTHA A. ROZELLE, The Rozelle Group, Ltd., Phoenix, Arizona
SARAH SLAUGHTER, Massachusetts Institute of Technology, Cambridge
RAE ZIMMERMAN, New York University, New York

Staff

RICHARD G. LITTLE, Director, Board on Infrastructure and the Constructed Environment
LYNDA L. STANLEY, Director, Federal Facilities Council
JOHN A. WALEWSKI, Program Officer
LORI DUPREE, Administrative Associate
AMANDA PICHA, Administrative Assistant

Acknowledgements

This report has been reviewed in draft form by individuals chosen for their diverse perspectives and knowledge of the subject matter, in accordance with procedures approved by the NRC Report Review Committee. The purpose of this independent review is to provide candid and critical comments that will assist the NRC in making this report as sound as possible and to ensure that it meets institutional standards for objectivity, evidence, and responsiveness to the study charge. The review comments and draft manuscript remain confidential to protect the integrity of the deliberative process. We wish to thank the following individuals for their participation in the review of this report:

Ms. Nancy Rutledge Connery, Woolwich, Maine
Dr. Joseph H. Golden, National Oceanic and Atmospheric Association
Dr. George W. Housner, California Institute of Technology
Dr. Dennis Mileti, University of Colorado
Dr. Dorothy A. Reed, University of Washington
Mr. Herbert Rothman, Weidlinger Associates
Dr. Robert H. Scanlan, Johns Hopkins University

Although these individuals provided constructive comments and suggestions, it must be emphasized that responsibility for the final content of the report rests with the authoring committee and the NRC.

ACKNOWLEDGEMENTS

Contents

Executive Summary

Executive Summary

The Idaho National Engineering and Environmental Laboratory (INEEL), through the U.S. Department of Energy (DOE), has proposed that a large-scale wind test facility (LSWTF) be constructed to study, in full-scale, the behavior of low-rise structures under simulated extreme wind conditions. To determine the need for, and potential benefits of, such a facility, the Idaho Operations Office of the DOE requested that the National Research Council (NRC) perform an independent assessment of the role and potential value of an LSWTF in the overall context of wind engineering research. The NRC established the Committee to Review the Need for a Large-scale Test Facility for Research on the Effects of Extreme Winds on Structures, under the auspices of the Board on Infrastructure and the Constructed Environment, to perform this assessment. This report conveys the results of the committee's deliberations as well as its findings and recommendations.

Data developed at large-scale would enhance our understanding of how structures, particularly light-frame structures, are affected by extreme winds (e.g., hurricanes, tornadoes, severe thunderstorms, and other events). Existing field data are based on observations and measurements of winds associated with the passage of frontal systems and a limited number of strong wind events. However, significant gaps exist in the meteorological data for severe wind events. Most data on structural loading has been derived from testing small-scale models in turbulent boundary-layer wind flow simulations; performance data have been collected from post-storm damage assessments and simplified tests of full-sized components. Mobile instrumentation systems have also been deployed in advance of storms to obtain data on the nature of extreme winds. New projects are being initiated by the National Oceanic and Atmospheric Administration (NOAA), the DOE, the National Institute of Standards and Technology, and several universities to gather wind data, measure structural loading, and observe structural performance during extreme wind events.

With a large-scale wind test facility, full-sized structures, such as site-built or manufactured housing and small commercial or industrial buildings, could be tested under a range of wind conditions in a controlled, repeatable environment. At this time, the United States has no facility specifically constructed for this purpose. The use of aeronautical testing facilities, such as the facilities operated by the National Aeronautics and Space Administration (NASA) at the Ames Research Center, has been discussed. However, additional study will be needed to determine if facilities of this type can be effectively used for large-scale structural research.

During the course of this study, the authoring committee was confronted by two difficult questions: (1) does the lack of a facility equate to a need for the facility? and (2) is need alone sufficient justification for the construction of a facility? These questions might not have engaged the committee at all if considerable resources were already available for wind engineering research and a coordinated national wind-hazard reduction program were in place. The committee found, however, that funding for research in wind engineering is only a few million dollars annually, and, despite some excellent programs and activities by government agencies and research institutions, research has not been strategically planned, coordinated, managed, or funded. Therefore, the committee raised a third question: would the benefits derived from information produced in an LSWTF justify the costs of producing that information? The committee's evaluation of the need and justification for an LSWTF was shaped by these realities.

The committee's evaluation is based on the logic tree shown in Figure ES-1.

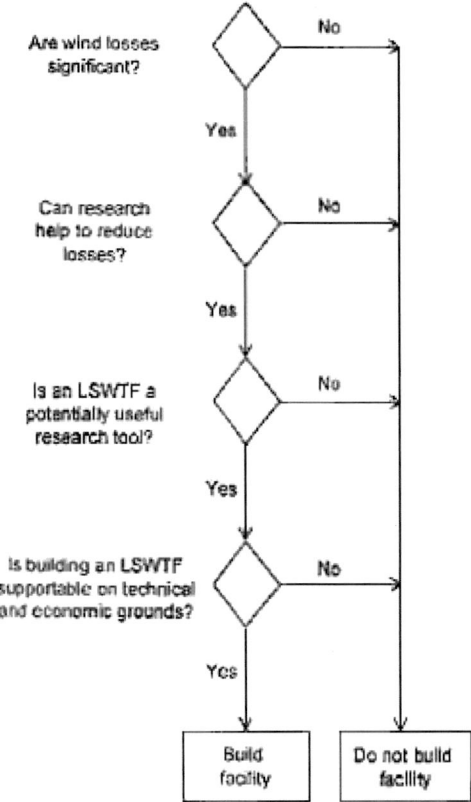

FIGURE ES-1
Logic tree used to assess the need for an LSWTF.

Based on the information available, as well as on the considerable experience of committee members in the field of wind-hazard reduction and large-scale structural research, the committee concluded that an LSWTF is unsupportable on both technical and economic grounds and recommends that the DOE not construct such a facility.

The committee believes that the interests of DOE, as well as the national interest, would be best served by DOE's participation in a cooperative effort involving federal government agencies, state and local governments, and research institutions, including universities and government laboratories. The cooperative effort should set research priorities, coordinate ongoing research, identify new opportunities, provide outreach to the building community and the general public, and implement new technologies and practices as they become available. To realize this program, the committee urges—in the strongest possible terms—that Congress consider designating funds for a coordinated national wind-hazard reduction program that encourages partnerships between federal, state, and local governments, private industry, the research community, and other interested stakeholders.

1

Introduction

One extreme wind event-Hurricane Andrew in 1992-inflicted the largest direct and indirect economic losses (~$25 billion) ever experienced by the United States as the result of a natural disaster (AAWE, 1997a). Although Hurricane Andrew was an extreme weather event, hurricanes, tornadoes, and storm surges in the United States cause, on average, several billion dollars in damage and claim hundreds of lives annually (Jones et al., 1995). The United States has made great improvements in its detection, warning, and reporting capabilities for major storms, increased awareness of the vulnerability of certain types of structures, and taken steps to mitigate damage. Despite these advances, the fatalities and damage from devastating storms has been growing, with individual dwellings and low-rise commercial and industrial structures bearing the brunt of the damage (NRC, 1985; Cermak, 1998).

In an effort to reduce these losses, particularly the loss of life, a small community of engineers and scientists has been conducting research for some decades into the nature of wind-structures interactions with the goal of improving the performance of non-engineered structures.[1] Although this research has led to some improvements in building codes and standards, materials selection, construction practices, and building inspection, major gaps remain in basic research and testing capabilities in wind engineering (Cermak, 1998).

Although several universities, private industries, and government laboratories have experimental and test facilities, no facility is capable of testing, to destruction, full-scale buildings of the type most prone to damage from extreme wind conditions (i.e., residences and non-engineered commercial buildings). Furthermore, even though large engineered structures have not suffered significant structural damage, the envelopes of these buildings are frequently seriously damaged by severe winds, causing considerable losses to contents and costly business interruptions.

The Idaho National Engineering and Environmental Laboratory (INEEL), through the U.S. Department of Energy (DOE), has proposed that a large-scale wind test facility (LSWTF) be constructed to determine the behavior of full-scale structures, including typical site-built and manufactured housing units, under extreme wind conditions in a controlled environment (INEEL, 1998). In order to determine the need for, and potential benefits of, such a facility, the Idaho Operations Office of the DOE requested that the National Research Council (NRC) perform an independent assessment of the role and potential value of an LSWTF in the overall context of research in wind engineering.

SCOPE OF THE STUDY

In response to that request, the NRC established the Committee to Review the Need for a Large-scale Test Facility for Research on the Effects of Extreme Winds on Structures under the

---

[1] For the purpose of this report, non-engineered structures are structures designed and constructed without the direct input of a registered, professional engineer. Essentially, all single-family homes are included in this category, as well as many multifamily homes and low-rise (one or two stories) commercial and industrial buildings.

auspices of the Board on Infrastructure and the Constructed Environment. The committee was asked to perform the following tasks:

- review the need for a large-scale, experimental, wind engineering facility
- identify the potential benefits of such a facility
- assess the priority of large-scale physical testing as a component of a national wind engineering research program

In addressing these tasks, the committee considered the following issues:

- the need for large-scale, experimental data for a better engineering/scientific understanding of the effects of extreme winds on non-engineered structures
- the benefits of generating data on extreme winds in a controlled environment as a complement to collected field data or to post-storm assessments
- the value of data produced by large-scale, full-system testing compared to small-scale or component testing
- the value of large-scale testing data (as compared to observational data) in the development and validation of computer simulations as a vehicle for (1) public education, (2) the validation of current building codes, and (3) improvements in the design of credible, standardized, small-scale or single-component experiments

ORGANIZATION OF THE STUDY

The 14 members of the study committee are renowned engineers and scientists with expertise in the following areas: wind-structure interactions, large-scale engineering research facilities, the performance of non-engineered structures, the characteristics of extreme winds, and wind-hazard reduction. Biographical information on the committee members is provided in Appendix A.

The committee met twice—once in December 1998 and once in January 1999. In light of the short time available to develop its findings and recommendations and issue a report, the committee drew heavily on the proceedings of three recent workshops and conferences on wind engineering (AAWE, 1997b; Marshall, 1995; O'Brien, 1996), two recent reports (AAWE, 1997a; NRC, 1993), and their own considerable experience. The committee also distributed a questionnaire to 75 researchers and practitioners in the fields of wind engineering, extreme wind events, and hazard mitigation. The questionnaires elicited 22 responses. The questionnaire, list of respondents, and synthesis of the responses are included in Appendix B. Although this report draws heavily on previously published work and responses to the questionnaire, the findings and recommendations were developed solely by the NRC committee that was specially appointed for this purpose.

ORGANIZATION OF THE REPORT

The succeeding chapters in this report address the committee's charge in the following manner. Chapter 2 contains a discussion of the technical aspects of an LSWTF and summarizes the committee's deliberations regarding the value of large-scale test data, wind-hazard research, uses and needs for large-scale testing, and the benefits and role of an LSWTF in wind engineering research. Chapter 3 is a discussion of economic considerations that the committee believes are relevant to an evaluation of an LSWTF. Chapter 4 contains the committee's findings and recommendations.

2

Technical Aspects of a Large-scale Wind Test Facility

INTRODUCTION

In the discussion that follows, the terms "large-scale" and "full-scale" are used in specific ways. "Large-scale" refers to structural models, or components of structures, with length-scale factors greater than, or equal to, 1:4. "Full-scale" refers to components, subsystems, or entire structures with length-scale factors of 1:1. Thus, full-scale testing is a subset of large-scale testing. Although the committee was not asked to evaluate a specific design, the type of LSWTF under consideration is assumed to be of the wind-tunnel type (as opposed to an actuator or pressure-chamber system) suitable for experimentation on large-scale components of structures, as well as testing (to failure) of full-scale selected structures (e.g., manufactured homes, residential buildings, and light commercial buildings). Experiments on these scales would require sustained wind speeds of 150 to 200 mph (~ 65 to 90 m/s), with a reasonable representation of atmospheric turbulence, over an area large enough to engulf a residential, single-family dwelling or other structure of similar size. The flow structure and size of the facility's wind stream would have to be sufficient to create a realistic flow around the structure and thereby generate appropriate and representative spatial and temporal variations of wind-induced pressures. At the present time, there are significant gaps in the meteorological data for severe wind events that would have to be filled before the design parameters and capability requirements for an LSWTF could be stipulated.

PREVIOUS ASSESSMENTS

Although there is general consensus in the wind engineering community about the need for large-scale data on the effects of extreme winds on structures, there is no consensus about the need for an LSWTF. The value of an LSWTF has been discussed in several assessments of research needs in wind engineering, including *Assessment of Wind Engineering Issues in the United States* (NRC, 1993); *Severe Windstorm Testing Workshop* (O'Brien, 1996); *Workshop on Large-scale Testing Needs in Wind Engineering* (AAWE, 1997b); and *Workshop on Research Needs in Wind Engineering* (Marshall, 1995), and was cited by several respondents to the committee's questionnaire. All of these assessments agreed that large-scale data are needed to improve structural performance and that an LSWTF could be a valuable tool for determining the effects of extreme winds on structures. These reports, however, also point out that other methods of data collection are available (e.g., full-scale field testing in natural wind) that may be able to answer many of the same questions. These reports concluded that the most effective research framework for wind-hazard reduction would be a combination of current methods of wind engineering research, such as full-scale field studies, wind tunnel and numerical modeling, component testing, and post-storm inspections. The reports emphasized that a coordinated national wind-hazard reduction program is necessary to mitigate wind-induced losses effectively, and they cautioned that an LSWTF alone would not provide answers to all outstanding questions in wind engineering (AAWE, 1997b; NRC, 1993). Some existing facilities in the United States and abroad might be modified for large-scale wind testing (AAWE, 1997a); another possibility is an international cooperative research program (NRC, 1993).

WIND-HAZARD RESEARCH

Minimizing the loss of life, property damage, and disruptions of economic activities from windstorms are primary objectives of wind engineering research. Consequently, any proposed national program or facility must be evaluated in light of whether it contributes significantly toward meeting these objectives. The Federal Emergency Management Agency (FEMA) and the insurance industry have both determined that significant improvements in the wind resistance of buildings will only be achieved when there is a demand for wind-resistant or hazard-resistant construction at the local and individual level (Cermak, 1998; FEMA, 1992). As a result, both FEMA and the insurance industry have embarked on pilot demonstration projects to highlight the benefits of hazard-resistant construction and other wind-hazard mitigation measures. Called Project Impact (FEMA, 1998) and the Show Case Communities (IBHS, 1998), these new projects have not yet demonstrated tangible results.

The research, engineering, and scientific communities have provided some of the technical underpinnings for reducing the vulnerability of buildings and other structures to wind damage. Significant work remains to be done in this area to ensure that key vulnerabilities of a particular structure are identified and that technically sound, cost-effective solutions are developed and implemented. Unfortunately, reducing vulnerability to wind-hazards is not just a question of developing appropriate technical solutions. First, wind-hazards are created by a variety of random events with large uncertainties in the magnitudes and characteristics of the winds. Second, the relevant government agencies and programs, as well as the construction industry, are fragmented. Third, implementation requires action by owners and the public, who may not consider hazard reduction a high priority. As a result, solving the wind-vulnerability problem will require coordinated work in scientific research, technology development, education, public policy, the behavioral sciences, and technology transfer.

In the past decade, several proposals have been put forward identifying the need for a national program of wind research, technology development, and education to address the technical needs for reducing losses associated with severe windstorms (NRC, 1993; Jones et al., 1995; Marshall, 1995; O'Brien, 1996; AAWE, 1997a). Despite these efforts, no national effort has been made to integrate wind research, technology development, and education into broader programs for natural hazard preparedness and disaster recovery (Cermak, 1997). Ultimately, losses associated with severe windstorms can only be significantly reduced if existing buildings and structures are modified and new buildings are designed, constructed, inspected, and maintained with wind resistance in mind.

An additional problem is the time it would take for the benefits of a coordinated plan to be observed. Only a small percentage of structures are replaced or added each year. Therefore, it would be many years before improvements in construction practices became prevalent. The adoption and implementation of remedial measures for existing structures is even more difficult to accomplish because the public often does not perceive a problem until a disastrous event occurs. The benefits and limitations of any single research facility must be carefully evaluated in light of the absence of coordinated action at the national level.

VALUE OF LARGE-SCALE TESTING

Testing of full-scale structures has been a part of wind engineering research for decades (Davenport, 1975), much of it associated with field measurements of wind characteristics, wind loads, and wind effects. These measurements have provided insight into the nature of various types of windstorms and benchmarks for evaluating analysis and design methods. Field studies continue to be an indispensable part of wind engineering research.

Data on the structure and characteristics of winds in severe windstorms are meager, however. Frequently, instrumentation, primary and backup power sources, and recording devices fail in severe windstorms, and the resultant data gaps leave large uncertainties about the magnitude and structure of winds in extreme events. The problem is complicated by the random structure and very large spatial gradients of wind, which makes it extremely difficult to characterize. For example, substantial differences in wind speeds and characteristics can be caused by changes in elevation and by averaging time associated with a particular observation, as well as the topography and roughness of the upwind terrain.

In an effort to reduce observational uncertainties in wind characteristics for extreme events, the National Oceanic and Atmospheric Administration (NOAA), the DOE, the National Institute of Standards and Technology (NIST), and several universities are attempting to measure wind magnitudes and wind characteristics in severe windstorms. New technologies are being employed, including new satellite imagery, airborne and ground-based Doppler radar (including two Doppler-on-wheels systems), wind profilers, Global Positioning System dropsondes, rapidly deployable trailers with anemometer masts, and new types of anemometers (Marks et al., 1998). All of these technologies were used during several recent hurricanes, which has led to considerable debate in the scientific, meteorological, and engineering communities regarding what is actually being observed and the implications of these observations. It will probably take several years of using these technologies before a coherent picture emerges.

Field studies of wind loads and wind effects on buildings have been even more limited (Eaton and Mayne, 1975; Hoxey and Richards, 1993; Levitan and Mehta, 1992a, 1992b; Marshall, 1975; Marshall, 1977; Robertson, 1991). No data are available on wind loads on buildings in the eye wall of a hurricane or in a tornado. No data on buildings subjected to thunderstorms and tropical storms have been reported in the literature. Experience with wind-tunnel model studies has shown that the gust structure of the wind plays an important role in the development of wind loads on structures. However, most of the existing field data on wind loads are limited to simple building shapes in open exposures subjected to winds generated by the passage of frontal systems rather than severe windstorms. The lack of knowledge about wind loading and structural response in severe windstorms is a significant impediment to establishing meaningful standards for structural systems and for improving structural performance.

A versatile, well conceived LSWTF could be used for a number of studies to identify and address data and knowledge gaps. Table 2-1 shows potential applications and technical capabilities that could be provided to the community of wind engineers and scientists. In addition, a number of other types of experiments might also be conducted, depending on costs and the availability of the facility.

For the purpose of reducing wind-hazards, an LSWTF would be most useful for conducting destructive experiments of large-scale structural systems, for fostering the development and validation of computational models, and for improving test methods. During the course of discussions and the review of responses to the questionnaire, the committee identified three topical investigations of buildings and structures that could be accomplished in an LSWTF: the performance of the building envelope, new construction techniques, and retrofitting technology.

- **Performance of the building envelope.** Economic assessments of damage following windstorms have shown that once a building envelope is compromised, losses increase dramatically (Cermak, 1998). An LSWTF could offer a practical approach to determining the wind speed at which the building envelope is compromised in a full-scale building.
- **Validation of construction techniques, practices, materials, and building code provisions.** Numerous remedial measures have the potential for improving the wind resistance of a building, and it is a relatively straightforward matter to test these measures at the component level. It is far more difficult, however, to assess the effectiveness of these measures in a full-scale system where their attributes interact synergistically. An LSWTF could provide an opportunity for assessing these measures under a range of controlled conditions thereby reducing the uncertainties about their effectiveness in severe winds. Significant advancements could be made in construction practices if the properties of a total building system could be evaluated in a full-scale turbulent wind flow representative of a hurricane, thunderstorm, or other extreme wind event.
- **Retrofitting techniques.** A comprehensive wind-hazard reduction program must include improvements to existing buildings. Retrofitting techniques can be tested as components of a system, but their value to the behavior of the full-scale building system can be determined only by testing a full-scale, complete system.

Destructive testing could include the following:

- **Testing of sheathing systems by applying realistic spatial and temporal variations of wind loads.** Current test methods apply loads uniformly over the surface of the specimen and have not included combined in-plane and out-of-plane loading.
- **Testing of the performance of the building envelope with emphasis on system performance relative to window and roof performance.** With current design criteria and construction practices, roof and wall systems may be more vulnerable to failure or water damage than protected windows and doors.
- **Testing of variations in internal pressures in a building with multiple rooms.** A breach of the building envelope, such as the failure of a window, can lead to pressurization of the building. Little is known about how pressurization is propagated throughout a building.

- **Validating retrofitting techniques in the context of overall building performance.** The benefits of a particular retrofitting measure or set of retrofitting measures could be ascertained, as well as whether the buildings will simply fail in another mode at a slightly higher wind speed.

TABLE 2-1 Technical Capabilities of a Large-Scale Wind Testing Facility (LSWTF)

Building Tests	Code Development and Validation	Other Applications	Instrumentation/Testing
High Reynolds Number testing of structural components	Validation of full-scale computational resistance models	Determining wind loads on floating offshore systems	Testing and calibration of new wind sensors
Water penetration experiments	Validation of computational fluid dynamics (CFD) models	Evaluating vehicle aerodynamics	Development of instrumentation concepts
Destructive testing of full-scale systems, including relationships to Saffir-Simpson Scale destruction categories	Validation of construction techniques, practices, materials, and building code provisions	Testing refinery systems (Reynolds Number)	Evaluation of wind generators
Sheathing system tests and evaluation that include spatial loads	Improving load/resistance characteristics	Tests of multiple steel stacks	Simplification of test methods
Strong room evaluations for residential structures	Validation of systemic retrofitting techniques		
Fatigue of elements and connections in a full-scale system	Development of damage fragility curves		
Window and roof system behavior relative to building envelope performance	Development of wind flow and energy use relationships		
Internal pressure distributions on internal walls and ceilings			
Damage sensitivity to wind speed characterization (peak gust, sustained wind)			
Windborne debris injection and transport			
Windborne debris impact phenomena			
Behavior of roof top appurtenances			
Behavior of roof edge attachment details			

- **Testing of windborne debris injection and transport.** It has been well established that severe windstorms generate and transport debris that can damage buildings and structures. Models of debris injection and subsequent transport are extremely sensitive to assumptions about wind speeds required to initiate the failure that produces the debris.
- **Testing of the performance of rooftop appurtenances.** Failure of mechanical system components and other rooftop appurtenances have caused significant damage to the interiors and contents of buildings.
- **Testing of the performance of porch roofs and roof overhangs.** Roof failures frequently originate at porches and roof overhang areas.

Uses of an LSWTF related to improving analytical models and simplified test methods could include the following:

- **Validation of full-scale computational resistance models.** Intense loading generally produces nonlinear structural behavior in certain components, connections, and at the system level. More realistic load modeling would result in more realistic modeling of the behavior of structural systems.
- **Validation of construction techniques, practices, materials, and building code provisions.** Rather than waiting for a storm to provide validation, it would be possible to create representative wind loading conditions in a controlled environment.
- **Realistic simulation of complex loading patterns and the response of the structural system to these loads.** Idealized loads specified in building code provisions and simplified analytic procedures sometimes lead to design requirements that are inconsistent with the observed performance of buildings in severe windstorms.
- **Development of improved component tests.** Many of the current tests for structural components and connections do not adequately reflect the actual physical processes at work in a severe windstorm.

Although this discussion has indicated that an LSWTF would be useful for wind engineering research, the rationale for establishing such a facility involves more than its capability to provide needed information. Many of the items listed above can be accomplished by other means (e.g., computational resistance models can be validated through full-scale measurements in natural wind or through comprehensive post-storm investigations). The low level of funding available for wind engineering research has been a major impediment to the development of new instrumentation, testing, and analytical technologies. It has also been a major impediment to the full and effective use of existing technologies to capture the variability of loads and resistance through wind-tunnel tests and component tests.

The committee noted that none of the major engineered structures in the world underwent full-scale testing to evaluate overall structural performance before it was built. With careful engineering, the wind resistance of low-rise residential and commercial structures could be dramatically improved. Given the current state of knowledge, a number of assumptions and considerable engineering judgments are necessary in the design of low-rise structures. In most cases, these assumptions and judgments lead to conservative designs. Thus, reducing the uncertainties could lead to economical designs more consistent with the actual level of risk. The real benefit of improved large-scale testing would be the savings and improved reliability of designs based on these investigations compared to engineered designs developed without the advantage of these experiments. Thus, the economic benefits of improved large-scale test methods, including an LSWTF, should be determined in terms of the savings expected compared to the cost of implementing better engineering design procedures, and not simply in terms of the potential savings over future construction using existing methods.

INEEL has proposed a pilot LSWTF to test manufactured housing. The committee believes that a more economical solution would be to deploy instrumented manufactured homes in the paths of hurricanes, surrounded by sufficient instrumentation to quantify the winds in the storm. The committee also believes that a large-scale pilot project is not a practical first step toward an LSWTF because the facility would have limited capabilities, could not provide the required data, and might preempt the development of a more general LSWTF.

ROLE OF A LARGE-SCALE WIND TEST FACILITY IN WIND ENGINEERING RESEARCH

Even with the modest funding currently available for wind engineering research, advances are being made in a number of areas, such as the characterization of wind fields and the evaluation of the performance of the building envelope (AAWE, 1997a). Two critical questions regarding the need for an LSWTF (as opposed to the desirability of having one) are whether it is uniquely capable of providing needed data and whether it can provide this information at lower cost than other alternatives. It may be that if the general level of funding for wind engineering research were significantly increased, much more could be accomplished in other ways, at lower cost, than by means of an LSWTF.

A variety of tools for research and development are available for determining the characteristics of wind-resistant structures, including analysis, numerical computation, wind-tunnel testing of small-scale models, wind-tunnel testing of large-scale or full-scale components, full-scale testing in the natural environment, and large-scale or full-scale testing of components and structures in simulated wind conditions under forces generated by actuators. Table 2-2 shows the scope and efficacy of a number of concepts for wind test structures. These tools have contributed to a growing understanding of how a wide range of structures, including tall buildings, low-rise commercial, industrial, and institutional buildings, residential buildings, and suspended-span bridges perform in high winds. Their potential for improving the economy and performance of structures of all types remains high.

However, this knowledge alone has not been sufficient for the widespread implementation of improved designs and construction methods. There are social, economic, and institutional barriers to the deployment of technological improvements that engineering research alone cannot address (Cermak, 1998). Therefore, although an LSWTF would be an additional tool that could potentially help to improve design and construction technology, the effective transfer of the information produced by such a facility into practice would have to overcome similar barriers.

Evaluating the efficacy of a wind engineering research method or facility requires first comparing its potential contributions with those of other experimental tools that could provide the same or equivalent information. To develop funding priorities, the relative costs of these tools must also be considered while recognizing that certain vital information may only be available from one form of experimentation, perhaps at considerable cost. Finally, the role of experimental investigations relative to other areas of needed wind engineering research must be considered, as well as how the greatest benefits can be achieved from the prudent investment of resources.

TABLE 2-2 Concepts for Wind Testing of Structures. Source: adapted from Cermak, 1998.

Concept Sufficiency: ✓ good to fair, — poor to not feasible	Loading Mechanism							
	Natural Wind		Artificial Wind		Modeled Wind		Synthetic Wind	
	Wind Speed (m/s)		Number of Fans		Boundary-Layer Wind Tunnel		Actuators	
Scale of Model Structure[a]	>45	<45	10–30[b]	1–2 mock-up tests[c]	new large	conventional	pressure chambers	dynamic wind load actuators
FULL-SCALE (1:1)								
Full Structure	✓	✓	✓	—	✓	—	—	✓[d]
Partial Structure	✓	—	✓	✓	✓	—	✓	✓[d]
LARGE-SCALE (1:1–1:4)								
Full Structure	✓	—	✓	—	✓	—	—	✓[d]
Partial Structure	—	—	✓	—	✓	✓	✓	✓[d]
MEDIUM SCALE (1:4–1:25)								
Full Structure	—	—	✓	—	✓	✓	—	✓[d]
Partial Structure	—	—	—	—	✓	✓	—	✓[d]
SMALL SCALE (1:25–1:250)								
Full Structure	—	—	—	—	✓[d]	✓[d]	—	—
Partial Structure	—	—	—	—	—	—	—	—

[a] one-story, one-family house
[b] creating a test section large enough to test large-scale structures (e.g., "Wall of Wind" in O'Brien, 1996)
[c] for example, a building glazing unit
[d] determine influence functions and compute structural system loads

Many different technical approaches have been brought to bear to improve the performance of the nation's building stock and infrastructure relative to wind loads. Continuing human and economic losses suggest that there is more work to be done in both the development and implementation of research results. There is a general consensus, however, that many of the results of current research have not been implemented effectively (Cermak, 1998).

The manufacturing sector needs to be involved in implementing research results because it supplies the large variety of materials and components that make up a constructed building. Engineers and contractors can only implement improvements if they have information on the performance of new products and materials. Because of the small market and difficulty of carrying out qualification tests on a limited budget, this information is not often developed. Therefore, a testing and certification mechanism should be established to assist manufacturers in qualifying proposed new items or concepts for improving the wind resistance of structures.

To date, the experimental focus in wind engineering has been in the use of wind tunnels, mostly boundary-layer wind tunnels (Cermak, 1995). Wind-tunnel facilities have provided a wealth of data and understanding about the nature of wind loads on a wide range of structures, but wind tunnels can only test models and cannot test causes of failure of structural elements. Although more needs to be done in this area, calibrations with (albeit limited) full-scale data suggest that the results are consistent with expected loads and pressures on real structures (Cermak, 1995). The results of wind-tunnel investigations, and supporting analytical and numerical computations, have led to significant improvements in building codes in the past two decades (Cermak, 1995). Related investigations have focused on evaluating the response of structural and nonstructural components (e.g., shear walls, roofing systems) to wind-induced loads, with testing performed frequently at large-scale, or even full-scale. Commercial testing—often proprietary—is also quite common. A number of complementary full-scale field investigations involving the use of natural environmental winds have also been performed. To date, these investigations have not included testing to failure.

The design of engineered structures has effectively incorporated aerodynamic characterizations obtained from wind-tunnel experiments, in some cases complemented by full-scale observations from the natural environment. For obvious reasons, no full-scale multistory building has ever been tested to failure under controlled conditions in an LSWTF. It is conceivable that at wind speed that would cause failure, experiments conducted on non-engineered structures in an LSWTF could provide information to improve current design practices. However, much can also be learned from analyses based on the results of component studies augmented by observations of failures in real events.

Structures designed to resist actual fluctuating wind loads would perform more predictably than structures designed according to current wind-load criteria and could possibly be less costly to build. The savings could be used to upgrade components of the building to further improve its overall performance. Analyses to failure of wood-frame homes, manufactured housing, and low-rise commercial structures, in conjunction with component testing, could help to determine their behavior leading to failure and improve their design. Experiments in an LSWTF could be used to validate computational results based on component and other tests for both steel buildings and wood-frame houses. However, validation is also possible from full-scale measurements (generally nondestructive) or, in a statistical sense, from detailed analyses of post-disaster damages.

It has been suggested that existing facilities in the United States or abroad could be modified for large-scale wind testing. The capabilities of at least one facility, the NASA Ames large-scale test facility, are described in the AAWE Report Workshop on Large-scale Testing Needs in Wind Engineering (AAWE, 1997b). Although this facility would have the capability to develop aerodynamic loading on structures as large as a manufactured house or a small residence, there would still be some significant difficulties in using it for wind-engineering investigations. The problems include the development of acceptably scaled turbulence and a significant concern that destructive testing would produce debris that could damage the wind tunnel or fans. Additional study would be required to determine if facilities of this type could be used for large-scale structural research.

PRIORITY OF A LARGE-SCALE WIND TEST FACILITY

Although this review was initiated at the request of DOE in response to a proposal by the INEEL, this committee was not asked to evaluate a specific proposal for an LSWTF. However, some important issues should be considered before any proposal is considered. First, funding for wind engineering research, technology transfer, and education in the United States has historically been about $4 million per year (AAWE, 1997a). Because a large-scale test facility would be only one of many tools available to the wind engineering community, and one with specific capabilities and limitations, it would be prudent not to spend a disproportionate amount of the available funds in any given year on the construction, maintenance, and operating expenses of an LSWTF. Figure 2-1 illustrates the committee's view of the relative importance of an LSWTF for wind-hazard reduction.

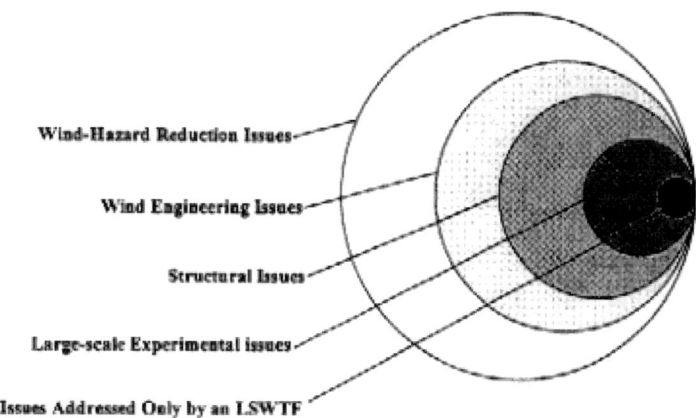

FIGURE 2-1
The Importance of an LSWTF in wind-hazard reduction.

Given that a large-scale test facility has the potential to be used in the ways already discussed, it is conceivable that such a facility could be a part of a well organized, well funded national wind-hazard reduction program at a later date. However, given the current state of wind engineering research, from the standpoint of overall funding and the capacity for technology deployment, the construction of an LSWTF at this time would be premature.

Before such a facility should be considered, a clear and objective plan for its use would have to be developed, describing exactly what capabilities the facility would include, the level of participation of the wind engineering research community in the research program, the specific questions that would be answered during the first few years of operation and at what cost, and the reasons these questions could not be answered more effectively, from both a technical and economic standpoint, by other means. Finally, there would have to be a clear understanding of how this facility and its research program and results would fit into a national wind-hazard reduction program.

3

Economic Considerations

Most major public expenditures of the magnitude required to design, construct, and operate an LSWTF are subjected to a rigorous economic analysis as an integral part of the process of deciding whether or not to proceed. The analysis might take the form of a benefit-cost study to determine if the expected benefits justify the costs or a return-on-investment calculation to identify less costly alternatives that could provide a similar quality and quantity of data. Although the committee was not asked to perform an economic analysis of an LSWTF, nor was it provided with the necessary data and resources to do so, the committee believes the following economics-related observations and conclusions are integrally related to the goals of this study.

The cost of constructing an LSWTF of the type described in Chapter 2 has been cited in various sources as $70 million to several hundred million dollars (INEEL, 1998; Philips, 1999; Haynes, 1999). The collective experience of the committee with large experimental facilities for both wind engineering and earthquake engineering research suggests that annual operating costs would average 5 to 10 percent of the construction cost ($5 million to $25 million). Individual experiments could cost $1 million or more. Although these costs are small compared to the dollar losses associated with wind-hazards on an annual basis, the committee doubts that the value of information produced in an LSWTF would justify them, given the current state of wind engineering research.

The decision to build an LSWTF carries with it an implicit commitment to fund the continuing operation of the facility. Operational costs are usually covered through user fees or direct appropriation. In light of the large potential operating and maintenance costs, the facility would have to operate on a nearly continuous basis to break even. However, it is not obvious to the committee that sufficient interest or capability exists, either in the government or in the private sector, to provide an adequate and sustainable source of funding for the operation of an LSWTF. In the areas of wind engineering and earthquake engineering, funds from private industry for research and experimentation on structures have been limited, in fact, practically insignificant, compared to funds provided by the federal government. Furthermore, there is no evidence to suggest that, even if the private sector were committed to supporting work at an LSWTF, private funds would be sufficient to sustain operations. Finally there is no guarantee that federal funds would be made available on an ongoing basis to support research at an LSWTF even if a national wind-hazard reduction program were established.

The committee believes that these economic realities must be considered in the decision process to build an LSWTF and offers the following conclusion. An LSWTF would be costly to build, operate, and maintain. Although valuable information could be produced at an LSWTF, the committee doubts that the value of this information would justify the cost of producing it. The committee is unable to identify a potential user base either in the United States or internationally to support operations on a user fee basis. There is no national wind-hazard reduction program, and funding for wind engineering research is limited (the committee believes inadequate). Redirecting existing funds to support LSWTF operations and maintenance would essentially put an end to ongoing research in the field and, therefore, cannot be advocated. Finally, building and operating an LSWTF using direct federal appropriations would be a poor use of government research funds.

In summary, the committee believes an LSWTF would be an extremely costly means of producing data that, as discussed in Chapter 2, would contribute in only a small way to reducing wind-related hazards and to advancing the general field of wind engineering research. Without quantitative economic data to evaluate, the committee was not able to address the potential benefits and costs of an LSWTF in detail. However, in the absence of any evidence to suggest that a favorable economic conclusion would result, and for the other reasons cites herein, the committee believes it would be uneconomical and inappropriate to construct an LSWTF.

4

Findings and Recommendations

Despite large and recurring losses from extreme wind events, no coordinated national program has been established for wind-hazard reduction. For many years, a small community of engineers and scientist has been conducting research into the nature of wind-structure interactions with the goal of improving the performance of structures most commonly damaged by extreme winds. The primary objective of this research is reducing, in an economically acceptable manner, the loss of life and property damage in future windstorms. Results of this research have led to improvements in building coded and standards, but they been applied unevenly in areas of high risk.

When the government formulates and implements a national wind-hazard reduction program, an LSWTF could be used to increase our knowledge and understanding of how residential and other low-rise structures behave in extreme winds. An LSWTF could create loadings for representative storm conditions and enable researchers to study their effects on building and other structures in a controllable, programmable, and repeatable environment. However, despite some favorable attributes, an LSWTF would also have technical limitations and economic drawbacks. Findings and recommendations of the committee are summarized below.

FINDINGS

Finding 1. Wind-structure interactions should be investigated at as close to full-scale as possible.

Although much valuable information can be developed from wind-tunnel tests of small-scale models, computational simulations, and other techniques, the complex interactions that occur within a total structural assembly subjected to extreme winds can only be determined by large-scale or full-scale experiments. In addition, data from large-scale experiments play an important role in validating numerical models and enhancing their credibility.

Finding 2. A variety of methods are available to investigate wind-structure interactions, and each method has both positive and negative aspects.

No single research or experimental method offers a means of reducing losses associated with extreme winds. Ultimately, losses will be significantly reduced only if existing buildings and other structures are remediated and changes are made in the design, construction, inspection, and maintenance of new structures. The committee believes that this can only be accomplished through a coordinated program of research, education, and technology transfer that integrates available and new knowledge from engineering and the physical and social sciences.

Finding 3. An LSWTF has the potential to perform experiments under controlled and repeatable conditions —an option not readily available in experiments with natural wind. In addition, an LSWTF has the potential, as a demonstration medium, for increasing public awareness of building performance in high winds.

Repeatable experiments at large-scale in a controlled environment cannot be performed in natural winds. This characteristic sets an LSWTF apart from other experimental method and is the reason the committee believes there may ultimately be a place for an LSWTF in a comprehensive national program for wind-hazard reduction. The ability to demonstrate building performance, including failure, might be useful to focus public interest on the need for mitigation.

Finding 4. Alternatives to an LSWTF are available that could potentially provide much of the same data.

A variety of tools for research on wind-resistant structures are available, including analysis, numerical computation, wind-tunnel testing of small-scale models, wind-tunnel testing of large-scale of full-scale components, full-scale testing in the natural environment, and large-scale or full-scale testing of components and structures under forces generated by actuators simulating wind action.

Finding 5. An LSWTF should only be built if it is part of a national wind-hazard reduction program. It should not precede the program or be considered a substitute for it.

Although an LSWTF could play a role in expanding knowledge, improving current practices, and providing demonstrations, experiments conducted at an LSWTF would be only one component of a national wind-hazard reduction program. In that context, it is clearly important to ensure that the costs of building, operating, and maintaining an LSWTF do not preclude the development and application of other more or equally effective facilities, tools, and techniques.

Finding 6. An LSWTF is an extremely costly method of producing data that will address only a small fraction of the issues in wind-hazard reduction.

The development of an LSWTF should not proceed until such time as the acquisition and operating costs of the facility represent a small fraction of the funds expended on wind-hazard reduction. Alternative methods already exist or can be developed that will provide much more cost-effective solutions for obtaining data and providing demonstrations. In addition, a facility would uniquely address only a small fraction of the pertinent issues in, and impediments to, wind-hazard reduction.

Finding 7. Construction of an LSWTF is premature.

Before an LSWTF could be properly designed and operated, advances would have to be made in understanding the characteristics of extreme winds. In addition, the barriers to the deployment of new technologies and practices for wind-hazard reduction cannot be overcome by engineering research alone. In the absence of a national wind-hazard reduction program, the results produced at an LSWTF are likely to be slow to find their way into practice.

Finding 8. Despite recurring and escalating losses from extreme winds, no coordinated national program for wind-hazard reduction currently exists.

A number of government agencies have wind-related mitigation and research programs including: FEMA, NOAA, NIST, the National Science Foundation, and DOE. Recently, the subcommittee on Natural Disaster Reduction of the National Science and Technology Council has been actively engaged in developing a national natural disaster reduction plan. However, there is still a lack of leadership, focus, and coordination of wind-hazard mitigation activities across all agencies, and funding for research and development specifically targeting wind-hazard reduction issues is insufficient.

RECOMMENDATIONS

Recommendation 1. The U.S. Department of Energy should not proceed with a large-scale wind test facility.

Recommendation 2. The federal government should coordinate existing federal activities and develop, in conjunction with state and local governments, private industry, the research community, and other interested stakeholder groups, a national wind-hazard reduction program. Congress should consider designating sufficient funds to establish and support a national program of this nature.

References

AAWE (American Association for Wind Engineering). 1997a. Wind Engineering: New Opportunities to Reduce Wind-Hazard Losses and Improve the Quality of Life in the USA. Notre Dame, Ind.: AAWE.

AAWE. 1997b. Workshop on Large-Scale Testing Needs in Wind Engineering sponsored by National Science Foundation and the National Institute of Standards and Technology. Washington, D.C.: AAWE.

Cermak, J.E. 1995. Development of Wind Tunnels for Physical Modeling of the Atmospheric Boundary Layer (ABL). A State of the Art in Wind Engineering. Pp. 1–25 in the 9th International Conference on Wind Engineering. London, U.K.: New Age International Publishers Limited.

Cermak, J.E. 1997. Prospectus for a National Wind Science and Engineering Program (NWSERP). Prepared as basis for discussion at AAWE meeting of December 7–8, 1997, at Johns Hopkins University, Baltimore, Maryland.

Cermak, J.E. 1998. Pp. 335–352, Wind Damage Mitigation: Wind Engineering Challenges, in Wind Effects on Buildings and Structures. Rotterdam: Balkema.

Devenport, A.G. 1975. Perspectives on the full-scale measurements of wind effects. Journal of Industrial Aerodynamics 1(1): 33–47.

Eaton, K.J., and J.R. Mayne. 1975. The measurement of wind pressures on two-story houses at Aylesbury. Journal of Industrial Aerodynamics 1(1): 67–109.

FEMA (Federal Emergency Management Agency). 1992. Building Performance: Hurricane Andrew in Florida. Washington, D.C.: FEMA.

FEMA. 1998. Project Impact Guidebook. Project Impact: Building a Disaster-Resistant Community. Internet: http://www.fema.gov/impact/guidebk.htm

Haynes, V.D. 1999. Idaho Site: Engineers to Test How Storms Affect Buildings. Chicago Tribune, January 3, 1999.

Hoxey, R.P., and P.J. Richards. 1993. Flow patterns and pressure fields around a full-scale building. Journal of Wind Engineering and Industrial Aerodynamics. 50(1/3): 203–212.

IBHS (Institute for Business and Homes Safety). 1998. Brochure: IBHS Showcase Communities. Boston, Mass.: IBHS.

REFERENCES

INEEL (Idaho National Engineering and Environmental Laboratory). 1998. Overview of INEEL, presentation before the committee, December 7, 1998, Washington, D.C.

Jones, N.P., D.A. Reed, and J.E. Cermak. 1995. National wind-hazards reduction program, ASCE Journal of Professional Issues in Engineering Education and Practice 121(1): 41–46.

Levitan, M.L., and K.C. Mehta 1992a. Texas Tech field experiments for wind loads. Part I: Buildings and pressure measurement system. Journal of Wind Engineering and Industrial Aerodynamics 41–44: 1565–1576.

Levitan, M.L., and K.C. Mehta. 1992b. Texas Tech field experiments for wind loads. Part II: Meteorological instrumentation and terrain parameters. Journal of Wind Engineering and Industrial Aerodynamics 41–44: 1577–1588.

Marks, F.D., Jr., L.K. Shay, G. Barnes, P. Black, M. DeMaria, B. McCaul, J. Molinari, M. Montgomery, M. Powell, B. Tuleya, G. Tripoli, L. Xie, and R. Zehr. 1998. Landfalling tropical cyclones: forecast problems and associated research opportunities. Bulletin of the American Meteorological Society 79(2): 305–323.

Marshall, R.D. 1975. A study of wind pressures on a single-family dwelling in model and full-scale. Journal of Industrial Aerodynamics 1(2): 177–200.

Marshall, R.D. 1977. The Measurement of Wind Loads on a Full-Scale Mobile Home. National Bureau of Standards (currently National Institute of Standards and Technology). Report NBSIR 77–1289. Gaithersburg, Md.: National Bureau of Standards.

Marshall, R.D. (ed.) 1995. Proceedings: Workshop on Research Needs in Wind Engineering. Gaithersburg, Md.: National Institute of Standards and Technology.

NRC (National Research Council). 1985. Hurricane Iwa, Alicia, and Diana — Common Themes. Committee on Natural Disasters. Washington, D.C.: National Academy Press.

NRC. 1993. Wind and the Built Environment: U.S. Needs in Wind Engineering and Hazard Mitigation. Washington, D.C.: National Academy Press.

O'Brien, C.C. (ed.) 1996. Full-Scale Structural Testing for Severe Wind: Proceedings of the INEL Severe Windstorm Testing Workshop. Idaho Falls, Idaho: Idaho National Engineering and Environmental Laboratory.

Philips, W.G. 1999. Preparing for Disaster. Popular Science 254(1): 39.

Robertson, A.P. 1991. Effect of eaves detail on wind pressures over an industrial building. Journal of Wind Engineering and Industrial Aerodynamics 38(2–3): 325–333.

REFERENCES

Appendixes

APPENDIXES

A

Biographies of Committee Members

Jack E. Cermak (NAE) chair, is University Distinguished Professor, Fluid Mechanics and Wind Engineering, Engineering Research Center, Colorado State University. Dr. Cermak specializes in teaching and research related to environmental science, aerodynamics, engineering mechanics, meteorology, and fluid mechanics. He is the recipient of many awards and honors including: North Atlantic Treaty Organization Post-doctoral Fellow at Cambridge University; Aeronautics and Astronautics Award for Distinguished Leadership in Aerospace Engineering, American Institute for Aeronautics and Astronautics; American Society of Mechanical Engineers (ASME) Freeman Scholar; Sigma Xi National Lecturer; lecturer for Southwest Mechanics Lecture Series; member of Colorado's Governor's Science and Technology Advisory Council; ASME National Distinguished Lecturer; Senior Research Award, American Society of Engineering Education; honorary member, American Society of Civil Engineers (ASCE); and national honor member, Chi Epsilon, the National Civil Engineering Honor Society. He has authored or co-authored more than 650 papers and reports and is editor or reviewer for a number of publications, including Mechanics Research Communications and the Journal of Wind Engineering and Industrial Aerodynamics. He founded the Fluid Mechanics and Wind Engineering Program at Colorado State University, as well as the Fluid Mechanics and Diffusion Laboratory, which was awarded the Outstanding Engineering Achievement Award from the National Society of Professional Engineers. Dr. Cermak was elected to membership in the National Academy of Engineering in recognition of his pioneering development of boundary-layer wind tunnels and served on the National Research Council Committee on Natural Disasters. He earned B.S. and M.S. degrees in civil engineering from Colorado State University and a Ph.D. in engineering mechanics from Cornell University.

Alan Davenport (NAE) is a professor of civil engineering at the University of Western Ontario. His research interests include aerodynamics, meteorology, environmental loads, structural dynamics, and earthquake loading. Dr. Davenport has pioneered the application of boundary-layer wind tunnels to the design of wind-sensitive structures, the description of urban wind climates, and other problems involving the action of wind. He was the founder of the Boundary Layer Wind Tunnel Laboratory and has been the director since it was established. Dr. Davenport has authored more than 200 papers and has lectured around the world. He has received numerous honors and awards including: Alfred Noble Prize; Gzowski Medal and Duggan Medal and Prize, Engineering Institute of Canada; Golden Plate Award, American Academy of Achievement; International Award of Merit in Structural Engineering, International Association of Bridge and Structural Engineering; Hellmuth Prize, University of Western Ontario; and the Canada Gold Medal for Science and Engineering. He has also received nine honorary degrees. He was elected to the Royal Society of Canada, the Fellowship of Engineering in England, and is a founding member and past president of the Canadian Academy of Engineering. He has been a consultant on the design of the World Trade Center, the Sears Building, and the Ting Kau Bridge in Hong Kong. Dr. Davenport received his B.A. and M.A. in mechanical sciences from Cambridge University, his M.A.Sc. in civil engineering from the University of Toronto, and his Ph.D. in civil engineering from the University of Bristol.

Michael P. Gaus is research professor of civil engineering at the State University of New York at Buffalo. Dr. Gaus specializes in teaching and research in the areas of earthquake, wind, and natural hazard engineering; dynamic response of structures to wind; the performance of civil engineering materials; and computer methods in structural analysis and design. He is the current president of the American Association for Wind Engineering. Dr. Gaus is the recipient of several awards and honors including: the Meritorious Service Medal, National Science Foundation; Award for Outstanding Contributions to Engineering, School of Engineering and Applied Sciences, George Washington University; Award for Outstanding Contributions to Wind Engineering, Wind Engineering Research Council. He has served on a number of committees at the ASCE, the National Academy of Engineering, the ASME, and the Earthquake Engineering Research Institute. Dr. Gaus has held positions at a number of universities, consulting firms, and the National Science Foundation, where he worked on the development of research activities in natural hazard engineering, including wind, flood, large-scale earth movements, drought, and expanding and shrinking soil hazards. Dr. Gaus received his B.S., M.S., and Ph.D. in civil engineering and theoretical and applied mechanics from the University of Illinois at Champaign-Urbana.

Stephen R. Hoover is a senior fire protection consultant with Kemper/NATLSCO, a Kemper Insurance-owned consulting firm. Mr. Hoover was a field engineer, account engineer, engineering supervisor, and staff engineer for the American Protection Insurance Company (a Kemper Company) before becoming a part of Kemper/NATLSCO. He has been involved with a number of committees including: Built-up Roofing Committee, American Society for Testing Materials; Uplift Testing Committee, ASTM; Committee for the Study of Hail Damage to Shingles, Insurance Institute for Property Loss Reduction; Rubber Tire Storage (chair), National Fire Protection Association (NFPA); Automatic Sprinkler (secretary), NFPA; and Inspection, Testing, Maintenance of Water Based Systems, NFPA. Mr. Hoover has attended several seminars on roofing technology at the University of Wisconsin and a wind engineering seminar at Texas Tech University. He has written several articles on roofing technology for REPORT, Plant Engineering, and Construction Specifier magazines. Mr. Hoover has taught roofing technology in Kemper education classes to both Kemper engineers and clients for 20 years. He has written all of the roofing, windstorm, snow load, and ponding portions of the NATLSCO Technical Reference Manual. Mr. Hoover received his B.S. in civil engineering from Indiana Institute of Technology.

Nicholas P. Jones is professor of civil engineering at the Johns Hopkins University. His research and teaching focuses on structural dynamics, system identification, flow-induced vibration, and wind and earthquake engineering. He co-founded an experimental research program on aeroelasticity and aerodynamics of civil engineering structures using the Corrsin wind tunnel at Johns Hopkins University (JHU). Dr. Jones has received numerous honors and awards including: George Owen Teaching Award, JHU; 1998 Maryland Young Engineer of the

Year, Maryland Engineers Week Council; National Science Foundation Presidential Young Investigator Award; Robert Pond Teaching Award, JHU; Huber Research Prize, ASCE; invited keynote speaker at the symposium in Kobe, Japan, inaugurating the opening of the Akashi-Kaikyo Bridge and at the International Symposium, "Advances in Bridge Aerodynamics, Ship Collision Analysis, and Operation and Maintenance," commemorating the opening of the East Belt Bridge in Denmark. He is incoming editor for the Journal of Wind Engineering and Industrial Aerodynamics, on the board of directors of the American Association for Wind Engineering, and recently chaired the 8th U.S. National Conference on Wind Engineering. Dr. Jones received his B.E. from University of Auckland, New Zealand, and his M.S. and Ph.D. degrees in civil engineering from the California Institute of Technology.

Ahsan Kareem is professor of civil engineering and geological sciences at the University of Notre Dame. Dr. Kareem specializes in research and teaching in probabilistic structural dynamics, fluid-structure interactions, and design of structures to resist natural hazards, including wind, waves, and earthquakes. Dr. Kareem is the recipient of numerous honors and awards including: 1998 Achievement in Academia Award, College of Engineering, Colorado State University; 1997 Engineering Award, National Hurricane Conference for Contributions to the Development of ASCE 7–95; Presidential Young Investigator Award, The White House Office of Science and Technology Policy/National Science Foundation; Halliburton Young Faculty Research Excellence Award, University of Houston; Martin Minta Award, American Institute for Aeronautics and Astronautics. He has also been the chairman of several committees including: Committee on Wind Effects/STD-Dynamics Effects, ASCE; Task Committee on Damping System/Wind Effects/STD-Dynamic Effects, ASCE; and Probabilistic Methods Committee, Engineering Mechanics Division, ASCE. He was a member of the National Research Council (NRC) Panel for Wind Division, ASCE Panel for Assessment of Wind Engineering Issues in the United States, and NRC Committee on Natural Disasters. Dr. Kareem served as a member of the board of directors on the Wind Engineering Research Council and is the immediate past president of the American Association for Wind Engineering. He is editor-in-chief, North and South American Wind and Structures; and associate editor of the Journal of Engineering Mechanics, ASCE. In addition, Dr. Kareem is a member of the following publications: Probabilistic Engineering Mechanics; Journal of Wind Engineering and Industrial Aerodynamics; Structural Safety, Engineering Structures, and Applied Ocean Research. He has served as a consultant to the United Nations Development Program and as a senior consultant to the oil, design, and insurance industries. He received his B.S. in civil engineering from West Pakistan University of Engineering and Technology, his M.S. in civil engineering from the University of Hawaii, and his Ph.D. in civil engineering from Colorado State University.

Richard Kristie is a consultant with Wiss, Janney, Elstner Associates in Northbrook, Illinois. Mr. Kristie is a licensed structural engineer and a licensed professional engineer in Illinois. He has worked on a number of large structural design and analysis projects and on the development of structural analysis software. He has specialized in investigations involving a variety of structure types and component systems including: wood structures, wood truss roof systems, steel structures with corrosion problems, fire damaged structures, and plaza waterproofing systems. Mr. Kristie performed investigations of more than 60 residential structures in south Florida that were damaged during Hurricane Andrew. Mr. Kristie co-authored a paper on plate-connected wood trusses presented at an international conference on timber engineering and was lead author of a paper on wood bowstring trusses published in the ASCE Practice Periodical on Structural Design and Construction. Mr. Kristie received his B.S. in civil engineering from the University of Illinois, Champaign-Urbana.

William F. Marcuson, III, (NAE) is the director of the Geotechnical Laboratory at Waterways Experiment Station of the U.S. Army Corps of Engineers. His areas of research expertise include dams, earthquake engineering, geotechnical engineering, and soil and rock mechanics and testing. He has been a member of several NRC committees, including the Advisory Panel for a National Earthquake Engineering Experimental Facility Study and the Workshop on Liquefaction. He is a member of many professional organizations including: American Society of Civil Engineers; International Society of Soil Mechanics and Foundation Engineering; American Society of Testing and Materials; Society of American Military Engineers; and the Earthquake Engineering Research Institute. In addition, Dr. Marcuson is a fellow of the Institution of Civil Engineers, a chartered engineer in England, and a licensed professional engineer in South Carolina and Mississippi. Dr. Marcuson received his B.S., M.S., and Ph.D. in civil engineering from the Citadel, Michigan State University, and North Carolina State University, respectively.

Joseph E. Minor is research professor at the University of Missouri-Rolla and a private consulting engineer. Dr. Minor is recognized internationally in the fields of wind engineering, window glass design practice, and natural hazards research. Special areas of expertise include wind-structure interaction phenomena, effects of tornadoes and hurricanes on buildings, performance of window glass and curtain wall systems, building code provisions for wind effects, the economics of wind-resistant construction, and the impact of natural hazards on socio-economic systems. Dr. Minor is active on building code committees, industrial advisory boards, and professional society committees and as a consultant to government agencies, trade associations, and private organizations. He has lectured nationally and internationally on topics related to the integration of wind engineering research into professional practice and participates regularly in short courses and seminars related to the practice of wind engineering and window glass design practice. Dr. Minor is a member of many professional organizations including: ASCE, National Society of Professional Engineers, American Meteorological Society, and the Southern Building Code Congress International. Dr. Minor received his B.S. and M.S. degrees in civil engineering from Texas A&M University and his Ph.D. in civil engineering from Texas Tech University. He is a licensed professional engineer in Texas, Missouri, and Florida.

Joseph Penzien (NAE) is the chairman of International Civil Engineering Consultants and professor emeritus at the University of California at Berkeley. His expertise is in the fields of structural dynamics, structures, earthquake engineering, engineering mechanics, and offshore structures. He has been the recipient of numerous awards and honors including: North Atlantic Treaty Organization Senior Science Fellowship; Research Prize, ASCE; National Science Foundation Senior Science Fellowship; Silver Medal of Paris; Elected Fellow, American Academy of Mechanics; Nathan M. Newmark Medal, ASCE; Alfred M. Freudenthal Medal, ASCE; George W. Housner Medal, Earthquake Engineering Research Institute (EERI); Elected Honorary Member, EERI; The Berkeley Citation; Elected Honorary Member, ASCE. He has served on several NRC committees, including the Advisory Committee for the International Decade for Natural Disaster Reduction and the Advisory Panel for a National Earthquake

Engineering Experimental Facility Study. He has been a consultant to the United Nations Educational and Scientific Cultural Organization, State of California Attorney General's Office, and numerous engineering companies, research facilities, and government agencies worldwide. Dr. Penzien received his B.S. in civil engineering from the University of Washington and his Sc.D. from the Massachusetts Institute of Technology.

Mark D. Powell is a research meteorologist for the National Oceanic and Atmospheric Administration's (NOAA) Hurricane Research Division (HRD), located at the Atlantic Oceanographic and Meteorological Laboratory in Miami, Florida. At HRD, he has been active in microscale and mesoscale studies, concentrating on boundary-layer wind-structure in landfalling hurricanes and hurricane rain-band thermodynamics. Recently he has been active in the development of standards for the measurement of surface winds. He is currently leading a project on real-time surface wind analysis for eventual transfer to the National Hurricane Center as a forecasting tool for hurricane specialists. Dr. Powell has served as lead project scientist on NOAA P3 hurricane research flights, the Genesis of Atlantic Lows Experiment, and the Tropical Experiment in Mexico. He holds a certified consulting meteorologist designation from the American Meteorological Society. He has served on several committees including: Research Committee of the Interdepartmental Hurricane Conference, NRC Disaster Study Team on Hurricane Hugo's Landfall in the Mainland United States, Meteorology Subcommittee of the ASCE Task Committee on Wind Damage Investigation, and the U.S.-Japan Natural Disaster Task Committee on Wind-Hazards. He has served on the board of the American Association for Wind Engineering and is a member of the American Meteorological Society and the American Geophysical Union. He has published articles in several journals, including Journal of Geophysical Research, Monthly Weather Review, Weather and Forecasting, Wind Engineering and Industrial Aerodynamics, Bulletin of the American Meteorological Society, and Shore and Beach. Dr. Powell received his B.S. from the Florida State University, his M.S. from Pennsylvania State University, and his Ph.D. from Florida State University.

Timothy A. Reinhold is associate professor of civil engineering at Clemson University. Dr. Reinhold's areas of research and teaching interest include: wind effects on structures; structural dynamics; reliability engineering; scale modeling studies; fluid-structure interaction; structural analysis; and failure investigations. He is currently involved in wind-load studies for low-rise and specialty structures, including the resistance of structures to wind effects. Dr. Reinhold's research has included projects to: improve the simulation of wind loads on low-rise structures, investigate wind loads for coastal structures, investigate retrofitting for existing structures subjected to high winds, and investigate the feasibility of a full-scale wind test facility. Dr. Reinhold serves on the Wind Effects Committee, ASCE, the Southern Building Code Congress International, and the ASCE-7 Standard Wind Loads Subcommittee. Dr. Reinhold received his B.S., M.S., and Ph.D. degrees in engineering mechanics from Virginia Polytechnic Institute and State University.

Eleonora Sabadell is the director of the Natural and Technological Hazards Mitigation Program at the National Science Foundation. This program, in the Division of Civil and Mechanical Systems, supports research on the consequences of weather-related hazards on the built and natural environments. She has served on the NRC Panel on Water Resources Planning. Dr. Sabadell has represented the U.S. government in many international, bilateral, and United

Nations programs and conferences. She has worked with public and private organizations in Japan, India, Brazil, Mexico, Taiwan, Pakistan, People's Republic of China, Spain, Italy, and other countries. At the present time, she is a member of the Subcommittee on Natural Disaster Reduction of the National Science and Technology Council. Dr. Sabadell is the author and editor of articles, reports, and proceedings and a member of editorial boards and several professional associations. She received her degrees in chemical engineering from the National University of Buenos Aires, Argentina.

Emil Simiu is a National Institute of Standards and Technology (NIST) fellow and a research professor at the Johns Hopkins University. Dr. Simiu has conducted research at NIST's Building and Fire Research Laboratory on: dynamic loads induced on structures by wind, ocean waves, and earthquakes; structural dynamics; structural reliability; and chaotic and fluid-elastic responses. He is the co-author, with R.H. Scanlan, of "Wind Effects on Structures" (3rd ed., Wiley, 1996). Dr. Simiu has been a consultant to industry, government, and the World Bank. He is a past chairman of the ASCE Committees on Wind Effects, Dynamic Effects, and the Reliability of Offshore Structures and recipient of the Federal Engineer of the Year Award from the National Society of Professional Engineers and the Gold Medal, U.S. Department of Commerce. Dr. Simiu received his first degree from the Institute of Civil Engineering, Bucharest, his M.S. in applied mechanics from Brooklyn Polytechnic Institute, and his Ph.D. from Princeton University.

B

Questionnaire, Respondents, and Synthesis of Responses

Questionnaire

The National Research Council, through the Board on Infrastructure and the Constructed Environment, has been requested by the Department of Energy to assess the need for a wind testing facility capable of subjecting full-scale, non-engineered structures (such as homes and small commercial buildings) to extreme wind conditions. This assessment must be completed by March 1, 1999. In order to assist the panel conducting the assessment, we are soliciting the views of a broad segment of those concerned with the effects of extreme winds on non-engineered structures.

The task of the panel is to:

- review the need for a large-scale experimental wind engineering facility
- identify the potential benefits of such a facility
- assess the priority for large-scale physical testing as a component of a national wind engineering research program

To assist them in addressing their task, the panel requests your input on the following questions:

1) What is the need for large-scale experimental data in gaining scientific understanding of the effects of extreme wind events on non-engineered structures?
2) What are the benefits of generating data on extreme wind events in a controlled environment, rather than collecting field data in natural wind or performing post-storm inspections?
3) What is the value of data produced by large-scale, full-system testing vs. small-scale or component testing?
4) What is the value of large-scale testing to develop and validate computer simulations, as a vehicle for public education, to validate current building code prescriptive standards, and to aid in the design of credible standardized small-scale or single component tests?

5) What would be the cost of generating data in a facility capable of subjecting full-scale, non-engineered structures to extreme winds, relative to the costs of collecting data from full-scale tests in natural wind, small-scale or component testing, or performing post-storm inspections?
6) Given the relative costs of the various data collection methods and the relative value of the data each produces, which methods represent the most cost-effective ways of improving the scientific understanding of the effects of extreme winds on non-engineered structures?
7) Which industries would be the most likely to use a facility capable of testing full-scale structures in a controlled environment and to what extent are they likely to use it?

Your response to these questions may be as detailed and lengthy as you wish but please try to highlight your critical points. The panel will hold its first meeting in mid-November and it would be most helpful if you could respond by **October 30, 1998.**

Please respond by fax to **(202) 334-3370** or by email to **mporterf@nas.edu.**

Name:
Title:
Organization:
Address:
Phone:
Fax:
e.mail:

RESPONDENTS

Vince A. Amatucci
Senior Member of Technical Staff
Aerosciences and Compressible Fluid
Mechanics Department
Sandia National Laboratories
Albuquerque, New Mexico

Maurice Bazin
Deputy Director
Large Technical Facilities
Office National d'Etudes et de Recherches
Aerospatiales
Paris, France

Joseph Golden
Senior Meteorologist
National Oceanic and Atmospheric
Administration
Silver Spring, Maryland

Michael J. Griffin
Technical Manager
Associate
EQE International
St. Louis, Missouri

George Housner
Carl F. Braun Professor of Engineering,
Emeritus
California Institute of Technology
Pasadena, California

Bonnie Johnson
Director
Aerodynamic Laboratories
National Institute for Aviation Research
Wichita State University
Wichita, Kansas

Atul L. Khanduri
Senior Engineer
Risk Management Solutions, Inc.
Menlo Park, California
Richard D. Marshall

Retired
Building and Fire Research
Laboratory
National Institute of Standards and
Technology (NIST)
Gaithersburg, Maryland

Jorge L. Martinez
Director
Low Speed Wind Tunnel
Texas Engineering Experiment Station
Aerospace Engineering Division
Texas A&M University
College Station, Texas

Jim McDonald
Department Chair
Civil Engineering
Texas Tech University
Lubbock, Texas

Kishor Mehta
Director
Wind Engineering Research Center
Texas Tech University
Lubbock, Texas

Jim Merva
Technical Underwriting Director
St. Paul Fire and Marine Insurance
Company
Saint Paul, Minnesota

Eugene E. Niemi, Jr.
Professor
University of Massachusetts-Lowell
Lowell, Massachusetts

Mark Perry
Lead Projects Engineer
Lockheed Martin Low Speed Wind Tunnel
Smyrna, Georgia

Jon Peterka
Vice President
Cermak Peterka Petersen, Inc.
Wind Engineering Consultants
Fort Collins, Colorado

Emil Simiu
NIST Fellow
Building and Fire Research Laboratory
National Institute of Standards and Technology
Gaithersburg, Maryland

Dave Surry
Research Director
Boundary Layer Wind Tunnel Laboratory
University of Western Ontario
London, Ontario
Canada

Terry C. Taylor
Principal Consulting Engineer
Haag Engineering
Houston, Texas

Henry Tieleman
Professor Emeritus
Engineering Science and Mechanics
Virginia Polytechnic Institute and State University
Blacksburg, Virginia

Christian 0. Unanwa
Assistant Professor
South Carolina State University
Orangeburg, South Carolina

George R. Walker
Operations Director-Strategic Development
Aon Re Australia
Sydney, Australia

Pete Zell
Ames Research Center
National Aeronautics and Space Administration
Moffett Field, California

SYNTHESIS OF RESPONSES

Twenty-two people responded to the questionnaires. The respondents, in general, indicated that full-scale or large-scale testing is important and that a large-scale facility could be a useful tool in wind engineering research. There was no agreement among them on whether or not a large-scale facility was necessary to obtain important data or if other testing methods (e.g. full-scale testing in natural wind) could provide the same information. Several of them indicated that interdisciplinary, coordinated research will be necessary to mitigate wind-related losses and that no facility should be established except as part of a well conceived national program. The respondents highlighted many benefits of a facility capable of testing large-scale structures in a simulated extreme wind environment, but they expressed concerns about the capability of such a facility to simulate the characteristics of the natural wind, as well as the potential costs, both startup and maintenance costs, of such a facility. In addition, some respondents noted that here are many large wind tunnels in this country that, with modification, might provide badly needed data and that these options should be fully explored before significant funds are devoted to the construction of a new facility.

Below are summaries of the responses to each question.

1) What is the need for large-scale experimental data in gaining scientific understanding of the effects of extreme wind events on non-engineered structures?

In general, the respondents indicated that there is a need for large-scale experimental data to help the public understand the relationship between wind speeds and wind damage. The misrepresentation of wind speeds in past extreme wind events may have misled the public about the destructive power of extreme winds. In addition, large-scale data could be useful for calibrating and validating small-scale or component tests.

There was no consensus among the respondents as to whether or not a facility for research on the effects of extreme wind conditions on structures is necessary to obtain this data. Some respondents expressed concerns about the ability of such a facility to simulate wind flows and loading from small-scale vortices like tornadoes. Others expressed concerns about cost and the number of users for such a facility. Some equivalent data could be obtained more cheaply by other methods, such as measurements of wind loads, which can be made with existing facilities. However, no existing facility is capable of large-scale destructive testing, and there is a very low probability of destructive force winds hitting an instrumented structure in the field.

Some respondents felt that conclusions from an ensemble of full-scale studies were likely to be significantly more valuable than those reached from any individual experiment. They also stressed the importance of coordinated national research programs.

2) What are the benefits of generating data on extreme wind events in a controlled environment, rather than collecting field data in natural wind or performing post-storm inspections?

The respondents were in general agreement that a well planned, full-scale facility capable of capturing the characteristics of natural wind has some distinct advantages over collecting field data in natural wind or performing post-storm inspections. The most common benefit highlighted by the respondents was the potential ability for quick results through experimental control. With a full-scale facility, there would be no waiting for the "big one" to hit the instrumented structure. The level of control over wind velocity, temperature, barometric pressure, and other variables would be much greater in a wind simulation facility than in natural wind. Additional advantages include the ease of instrumenting and observing the behavior of the test structure. Another important benefit of a full-scale facility is the capability of repeating test conditions.

Many respondents again pointed out that no single data collection method would be adequate, and that an interdisciplinary approach would be necessary.

3) What is the value of data produced by large-scale, full-system testing vs. small-scale or component testing?

Although respondents highlighted many benefits of full-scale testing over small-scale or component testing, they also pointed out that tests at all scales have significant and complementary value and should be a part of an integrated national program. Full-scale testing can reveal some of the more subtle aspects of fluid mechanics and eliminate some difficult scaling issues, such as how to scale material properties. In addition, full-scale testing would allow for the determination of correct natural frequencies of structural systems and enable the study of aerodynamic and structural interactions and end or boundary fixity conditions between components that cannot be easily simulated by testing one piece at a time. The proposed facility would enable studies of progressive damage to failure so that the wind speeds associated with the onset of specific damage could be determined.

A full-scale testing facility could be good publicity and increase public awareness of the dangers posed by winds, provided the characteristics of extreme winds were adequately simulated and the costs of the facility were not so extreme that it would negatively affect public opinion about artificial destructive testing.

4) What is the value of large-scale testing to develop and validate computer simulations, as a vehicle for public education, to validate current building code prescriptive standards, and to aid in the design of credible standardized small-scale or single component tests?

The respondents indicated that it is clearly necessary to validate computer simulations, building codes, and small-scale or component tests. If a large-scale facility could properly model the natural wind, it might contribute to these validations. The question raised by many respondents, however, is whether or not there are more cost-effective ways to validate methods. It was suggested, for example, that computer simulations can be validated by field measurements and that code verification can be done with laboratory experiments.

The use of a large-scale facility as a vehicle for public education was also a point of debate among the respondents. While many believed that it would be educational and useful for people to see video footage of structures being blown apart in a facility, others argued that footage from actual events, during or after storms, sends a much stronger message than "fake" destruction. Others indicated that they felt there were equally effective and less expensive ways to educate the public.

5) What would be the cost of generating data in a facility capable of subjecting full-scale, non-engineered structures to extreme winds, relative to the costs of collecting data from full-scale tests in natural wind, small-scale or component testing, or performing post-storm inspections?

In general, the respondents seemed to be in agreement that a large-scale facility would be extremely expensive to build and operate. Many agreed that many other projects could be funded for the price of constructing a facility of this type. However, as one respondent pointed out, the costs of rebuilding non-engineered structures after a storm are also very large, and these costs should be considered when budgeting for wind engineering research. Even though the facility would be expensive, getting some data, especially failure data, through other means (e.g. full-scale testing to destruction in natural wind) would be virtually impossible. Paying large sums of money for unique, high-quality data may be appropriate and necessary for effectively mitigating wind-hazards.

Several respondents pointed out that scaling is difficult in small-scale (especially destructive) testing, but that small-scale and component tests have been beneficial in the past and would continue to be so. Large-scale, full-system testing may have advantages over component testing because the aerodynamic interactions between various components could be studied. One respondent pointed out the advantages of full-scale testing over post-storm inspections, which may not yield accurate data because they are subjective and because recording devices capable of accurately depicting structure-level wind conditions throughout the storm are not readily available. Given this current lack of ground-level wind data during extreme wind events, it might be premature to construct a simulation facility, especially given the fact that other methods (e.g. full-scale field and component testing) have been beneficial in the past.

6) Given the relative costs of the various data collection methods and the relative value of the data each produces, which methods represent the most cost-effective ways of improving the scientific understanding of the effects of extreme winds on non-engineered structures?

There was a wide range of opinions about which method of testing is most cost-effective. Several respondents indicated that they felt that a full-scale test facility would be the most cost-effective data collection method, while others argued that the enormous start-up and maintenance costs of such a facility would put it out of reach of anyone except the government. Some pointed out that post-storm surveys are a relatively inexpensive way of collecting data about existing structures and should be continued.

During severe storms, many structures are exposed to winds of similar magnitude simultaneously, and good comparisons can be made of different construction techniques. Other respondents indicated that small-scale tests have been valuable in the past and can continue to contribute to the knowledge base. Well planned small-scale testing with a few carefully executed full-scale or large-scale studies could greatly improve our understanding of wind effects on structures. They also pointed out that a number of facilities in the United States (e.g. wind tunnels at NASA Ames Research Center) might be adaptable for large-scale testing on wind effects on structures.

7) Which industries would be the most likely to use a facility capable of testing full-scale structures in a controlled environment and to what extent are they likely to use it?

Respondents suggested that the customer base would depend on the flexibility of the facility, the perceived realism of the wind simulation, and the cost per experiment. Serious concerns were raised by some respondents about whether or not anyone, except possibly the government, would have the financial resources to support full-scale or large-scale testing in such a facility. Possible customers that were suggested include:

- insurance industry
- government agencies
- construction industry
- prefabricated building industry
- educational institutions
- building code developers
- code enforcement authorities
- risk management companies
- roofing, component, and cladding companies

Acronyms

DOE	U.S. Department of Energy
FEMA	Federal Emergency Management Agency
INEEL	Idaho National Engineering and Environmental Laboratory
NASA	National Aeronautics and Space Administration
NIST	National Institute of Standards and Technology
NOAA	National Oceanic and Atmospheric Administration
NRC	National Research Council